Sharks

Thresher Sharks

by Julie Murray

Dash!
LEVELED READERS
An Imprint of Abdo Zoom • abdobooks.com
1

Level 1 – Beginning
Short and simple sentences with familiar words or patterns for children who are beginning to understand how letters and sounds go together.

Level 2 – Emerging
Longer words and sentences with more complex language patterns for readers who are practicing common words and letter sounds.

Level 3 – Transitional
More developed language and vocabulary for readers who are becoming more independent.

abdobooks.com

Published by Abdo Zoom, a division of ABDO, PO Box 398166, Minneapolis, Minnesota 55439.

Printed in the United States of America, North Mankato, Minnesota.
102019
012020

Photo Credits: Alamy, Seapics.com, Shutterstock
Production Contributors: Kenny Abdo, Jennie Forsberg, Grace Hansen, John Hansen
Design Contributors: Dorothy Toth, Neil Klinepier, Victoria Bates

Library of Congress Control Number: 2019941283

Publisher's Cataloging in Publication Data

Names: Murray, Julie, author.
Title: Thresher sharks / by Julie Murray
Description: Minneapolis, Minnesota : Abdo Zoom, 2020 | Series: Sharks | Includes online resources and index.
Identifiers: ISBN 9781532129230 (lib. bdg.) | ISBN 9781098220211 (ebook) | ISBN 9781098220709 (Read-to-Me ebook)
Subjects: LCSH: Thresher sharks--Juvenile literature. | Sharks--Juvenile literature. | Fish--Juvenile literature. | Ocean animals--Juvenile literature. | Top predators--Juvenile literature. | Carnivores-Juvenile literature.
Classification: DDC 597.33--dc23

Table of Contents

Thresher Sharks

Thresher sharks are found around the world. They mainly live in deep ocean waters.

Their backs are dark in color. Their bellies are lighter. They have spots near their tails.

They have **thick** bodies. They can be 20 feet (6.1 m) long!

Their tail fins make up much of their size. Their tails can be 10 feet (3 m) long!

Their tail fins are powerful. They can whip them 30 mph (48.3 kph)! This can **stun** or kill their **prey**.

Thresher sharks often hunt **schools** of fish. They swim circles around the fish.

They eat tuna and **mackerel**. They also like squid.

They can jump out of the water. Their tails **propel** them into the air.

They are not aggressive. They usually stay away from humans.

More Facts

- Thresher sharks are the only sharks to use their tails to hunt.
- They have small, sharp teeth. These help them catch and eat their **prey**.
- They migrate long distances. They spend summers in northern areas. They swim south in the winter.

Glossary

mackerel – a fish that has dark marks that look like waves on its back and a silver underside. They live in the Atlantic Ocean.

prey – an animal being hunted, caught, and eaten by another animal.

propel – to cause to move forward.

school – a large group of fish swimming together.

stun – to shock or cause to be unconscious.

thick – large from one end to the other.

Index

Online Resources

To learn more about thresher sharks, please visit **abdobooklinks.com** or scan this QR code. These links are routinely monitored and updated to provide the most current information available.